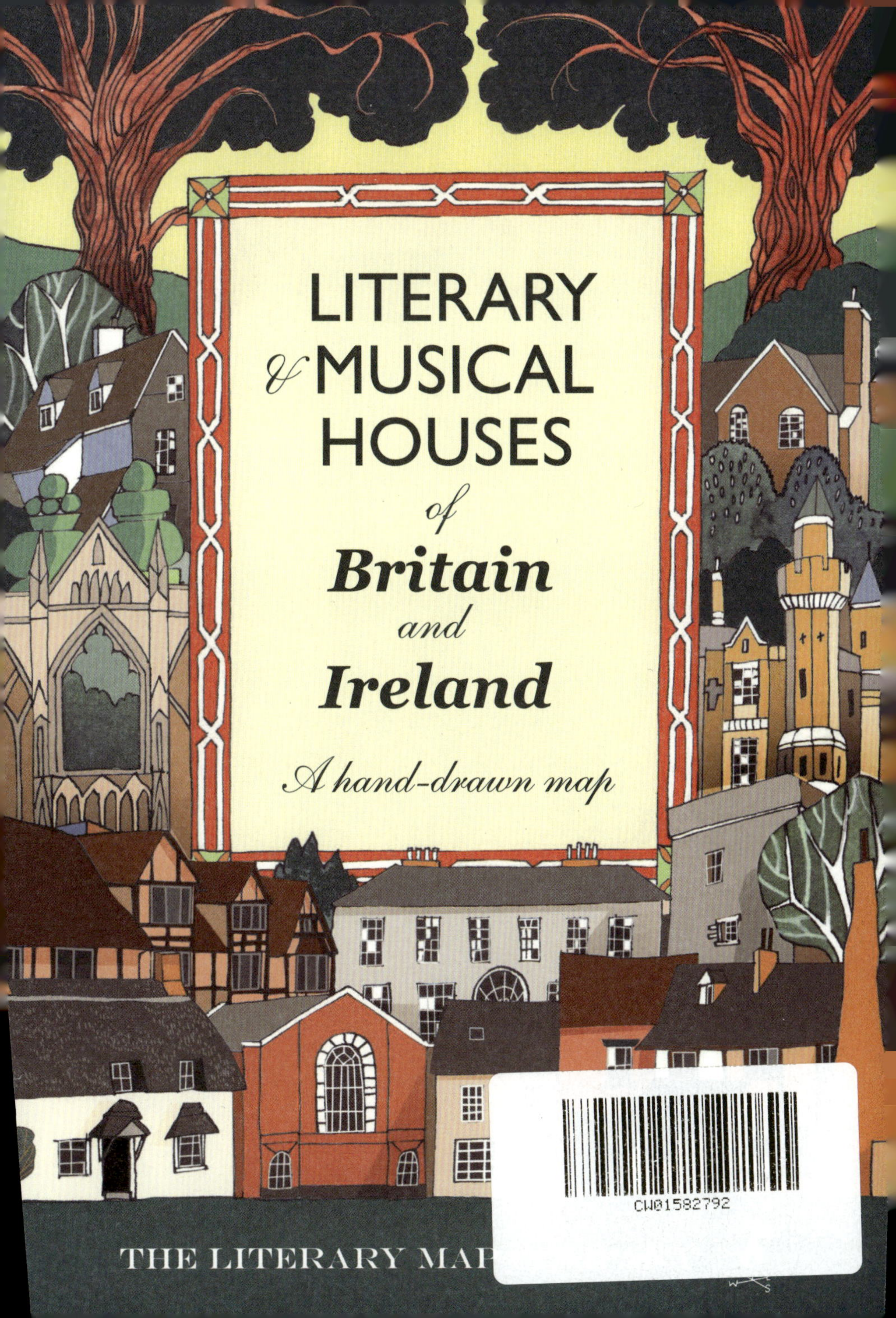

LITERARY & MUSICAL HOUSES of Britain and Ireland

A hand-drawn map

THE LITERARY MAP

Psychoanalysis simplifies life... It supplies the thread that leads a man out of the labyrinth of his own unconscious.
Sigmund Freud

No one is useless in this world who lightens the burdens of another.
Charles Dickens

When a man is tired of London, he is tired of life. For there is in London all that life can afford.
Samuel Johnson

Worcester
The Firs: Elgar's Birthplace
Stratford-upon-Avon
Shakespeare's Birthplace
Northampton
John Bunyan Museum
Hemingford Grey
Cambridge
(Britten and Pears) The Red House, Aldeburgh
Olney
Bedford
Cowper & Newton Museum
Milton Keynes
Luton
Stevenage
Knebworth House (Edward & Rosina Bulwer Lytton; Constance Lytton)
Ayot St Lawrence
Shaw's Corner
Cheltenham
Holst Birthplace Museum
Cotswolds
Oxford
(C.S.Lewis) The Kilns
(Disraeli) Hughenden
Roald Dahl Museum
Great Missenden
Lechlade
Kelmscott Manor (William Morris)
High Wycombe
Chalfont St Giles
Milton's Cottage
London
Walthamstow
William Morris Gallery
Strawberry Hill (H.Walpole)
Bexleyheath
The Red House (William Morris)
Gad's Hill Place (Dickens)
Down House
Downe
(Charles Darwin)
Cardiff
Norwegian Church (Roald Dahl)
Bristol
Beckford's Tower
Bath Jane Austen Centre
Guildford Museum (Lewis Carroll)
Knole (Vita & Eddy Sackville-West)
Broadstairs
Dickens House Museum
Chartwell (Churchill)
(Vaughan Williams) Leith Hill Place
Canterbury
Penshurst Place (Philip Sidney)
Royal Tunbridge Wells
Dover
Jane Austen's House
Alton
Chawton
Chawton House (Robinson, Austen Blackwell et al.)
Selborne
Gilbert White's House
Smallhythe Place (Christopher St John)
Nether Stowey
Coleridge Cottage
Taunton
Wilton House (Mary Herbert)
Salisbury
Burwash
Bateman's (Rudyard Kipling)
Rye
Lamb House (Henry James; E. F.Benson; Rumer Godden)
South Downs
Southampton
Charles Dickens Birthplace Museum
Lewes
Rodmell
Monk's House (Virginia Woolf)
New Forest
Portsmouth
Felpham
Brighton
Blake's Cottage
Hardy's Cottage
Higher Bockhampton
(T.E.Lawrence) Clouds Hill
Dorchester
Max Gate (Thomas Hardy)
Isle of Wight
Farringford (Tennyson)
Torquay
Greenway (Agatha Christie)
Hampstead Heath
Keats House
Hampstead
Freud
London

DUBLIN [D]

Dublin Writers Museum 18 Parnell Square N, D01 T3V8. 18th century mansion, celebrates literary Dublin through lives and works of, amongst others: **Swift**, **Sheridan**, **Shaw**, **Wilde**, **Yeats**, **Joyce** and **Beckett**. writersmuseum.com 01 8722077

James Joyce Centre 35 N Great George's St Dublin 1. Cultural Centre in Georgian townhouse, dedicated to the biography and works of influential avant-garde writer ***James Joyce*** (1882-1941). Works include *Ulysses* and *Dubliners* 01 878 8547 jamesjoyce.ie

Museum of Literature Ireland (**MoLI**) 86 St Stephen's Green, D02 XY43 Creative alliance of **University College Dublin** & the **National Library of Ireland** around historic Newman House gives access to Ireland's rich literary heritage, past to present. moli.ie 01 477 9811

Trinity College, ***Book of Kells*** & **Old Library** College Green, Dublin 2. Ireland's greatest cultural treasure, created c800AD probably on Iona, moved to, possibly finished in, Kells, Co.Meath; now housed in Trinity's beautiful 18th century library. 01 896 2320

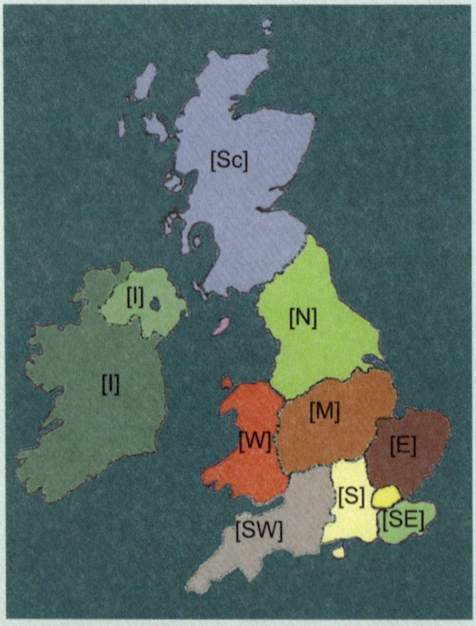

The Authors & Musicians

ap Gwilym [W]
Austen [S] [SW]
Barrie [Sc]
Beckett [D]
Beckford [SW]
Benson [SE]
Blackwell [S]
Blake [S]
Boston [E]
Boswell [Sc]
Britten [E]
Bronte [N]
Bulwer Lytton [E]
Bunyan [E]
Burns [Sc]
Byron [M]
Carlyle [Sc] [L]
Carroll [S]
Chaucer [L]
Christie [SW]
Churchill [SE]
Clare [E]
Coleridge [SW]
Cookson [N]
Cowper [E]
Dahl [S]
Darwin, Charles [GL]
Darwin, Erasmus [M]
Dickens [S] [SE] [L]
Disraeli [S]
du Maurier [SW]
Edgeworth [I]
Elfed [W]
Elgar [M]
Franklin [L]
Freud [L]
Gaskell [N]
Godden [SE]
Gore-Booth [I]
Gregory [I]
Griffiths [W]
Handel [L]
Hardy [SW]
Heaney [I]
Hendrix [L]
Herbert [SW]
Herriot [N]
Holst [SW]
James [SE]
Johnson [M] [L]
Joyce [I] [D]
Kavanagh [I]
Keats [L]
Kipling [SE]
Lawrence, D.H. [M]
Lawrence, T.E. [SW]
Ledwidge [I]
Lenin [L]
Lennon [N]
Lewis, C.S. [S]
Lister [N]
Lytton [E]
Mackay Brown [Sc]
Markievicz [I]
Martin [I]
Maxwell [Sc]
McCartney [N]
Milton [S]
Morgan [M]
Morris [GL] [SW]
Newton [E]
Nicholson [N]
O'Connor [I]
Owen [Sc]
Pankhurst [N]
Pears [E]
Potter [N]
Roberts [W]
Robinson [S]
Ruskin [N]
Sackville-West [SE]
Sassoon [Sc]
Scott [Sc]
Shakespeare [M]
Shaw [E] [D]
Sheridan [D]
Sidney [SE]
Somerville [I]
Sterne [N]
Stevenson [Sc]
StJohn/Marshall [SE]
Swift [D]
Tennyson [S]
Thomas [W]
Vaughan Williams[SE]
Walpole [GL]
Walton [M]
White [S]
Wilde [D]
Woolf [SE]
Wordsworth [N]
Wyn [W]
Wynne [W]
Yeats [I] [D]

Handel & Hendrix in London 25 Brook Street, Mayfair, London W1K 4HB 020 7495 1685 Two influential musicians and very different homes, separated by 200 years and a wall: opera and court composer of *The Messiah* ***George Frideric Handel*** (1685-1759) and ***Jimii Hendrix*** (1942-1970) American rock guitarist, singer/songwriter. handelhendrix.org 020 7495 1685

John Keats House 10 Keats Grove, Hampstead, London NW3 2RR Regency villa, home of major and much loved Romantic poet ***John Keats*** (1795-1821), where he wrote 'Ode to a Nightingale' and fell in love with 'the girl next door'. 020 7332 3868

Marx Memorial Library 37a Clerkenwell Green, London, EC1R 0DU 18th century building where ***Lenin*** (1870-1924) edited the Russian socialist newspaper *Iskra*. Tours. marx-memorial-library.org.uk 0207 253 1485

Westminster Abbey 20 Dean's Yard, London, SW1P 3PA Gothic abbey church. 'Poets' corner', burial and memorial place of writers since ***Geoffrey Chaucer'***s burial in 1400. westminster-abbey.org 020 7222 5152

IRELAND [I]

Drishane House Castletownshend, Skibbereen, Co. Cork P81 Y008 Classic Georgian home of the multi-talented ***Edith Somerville*** (1858-1949) who wrote mostly with her cousin ***Violet Martin*** as ***Somerville and Ross***, notably *The Experiences of an Irish R.M.* drishane.com Private house; visits only as specified on website.

Francis Ledwidge Museum Janeville, Slane, Co.Meath C15 DK82 19th century farm labourer's cottage, birthplace of First World War poet ***Francis Ledwidge*** (1887-1917) 041 9824544 francisledwidge.com

Frank O'Connor House 84 Douglas Street, Ballintemple, Cork. Vernacular 18th century house, birthplace of prolific writer and Irish nationalist ***Frank O'Connor/Michael O'Donovan*** (1903-66), best known for his short stories and memoirs. The house is now the base for the Munster Literature Centre. 021 4312955 munsterlit.ie

HomePlace 45 Main St, Bellaghy, Magherafelt BT45 8HT Purpose-built arts and literary centre celebrating the life and work of the poet and Nobel Laureate, ***Seamus Heaney*** (1939-2013) 028 7938 7444 seamusheaneyhome.com

James Joyce Tower and Museum Sandycove Point, Dún Laoghaire, Dublin, A96 FX33 19th century Martello tower, famous from the opening of ***Joyce***'s Ulysses. joycetower.ie

Kiltartan Gregory Museum Kiltartan Cross, Co. Galway, H91 C85C Former schoolhouse now a museum of local history and the Irish Literary Renaissance, in particular the parts played by Lady ***Augusta Gregory*** (1852-1932), Gaelic translator and co-founder of the Abbey Theatre, Dublin, and by ***W.B.Yeats***. ladygregoryyeatstrail.com

Lissadell House 19th century neo-classical-style country house, home of poet, artist and suffragist ***Eva Gore-Booth*** (1870-1926), close friend of ***W.B.Yeats***, and her sister, ***Constance Markievicz***, Irish freedom fighter and politician, first woman elected to the Westminster parliament. Private house. Tours lissadellhouse.com

Maria Edgeworth Centre Ballymahon Rd, Edgeworthstown, Co. Longford, N39 E3C8 1840s schoolhouse. Promotes the legacy of early realist Anglo-Irish educationalist and novelist ***Maria Edgeworth*** (1768-1849) and her family. 043 667 1801 edgeworthstown.net

Patrick Kavanagh Centre Candlefort, Inniskeen, Co. Monaghan, A91 KT70 Dedicated to the poet, novelist and critic ***Patrick Kavanagh*** (1904-67). Works include 'The Great Hunger' 042 937 8560 patrickkavanaghcentre.com

Thoor Ballylee Gort, Co. Galway, H91 D8F2 Hiberno-Norman castle and cottage, restored as a home by the poet ***W.B.Yeats*** (1865-1939); inspiration for poems such as 'The Tower'. yeatsthoorballylee.org

Yeats Building Hyde Bridge, Abbeyquarter North, Sligo. Arts and Crafts-style building, now a literary hub with exhibition on ***Yeats***' life. yeatssociety.com 071 914 2693

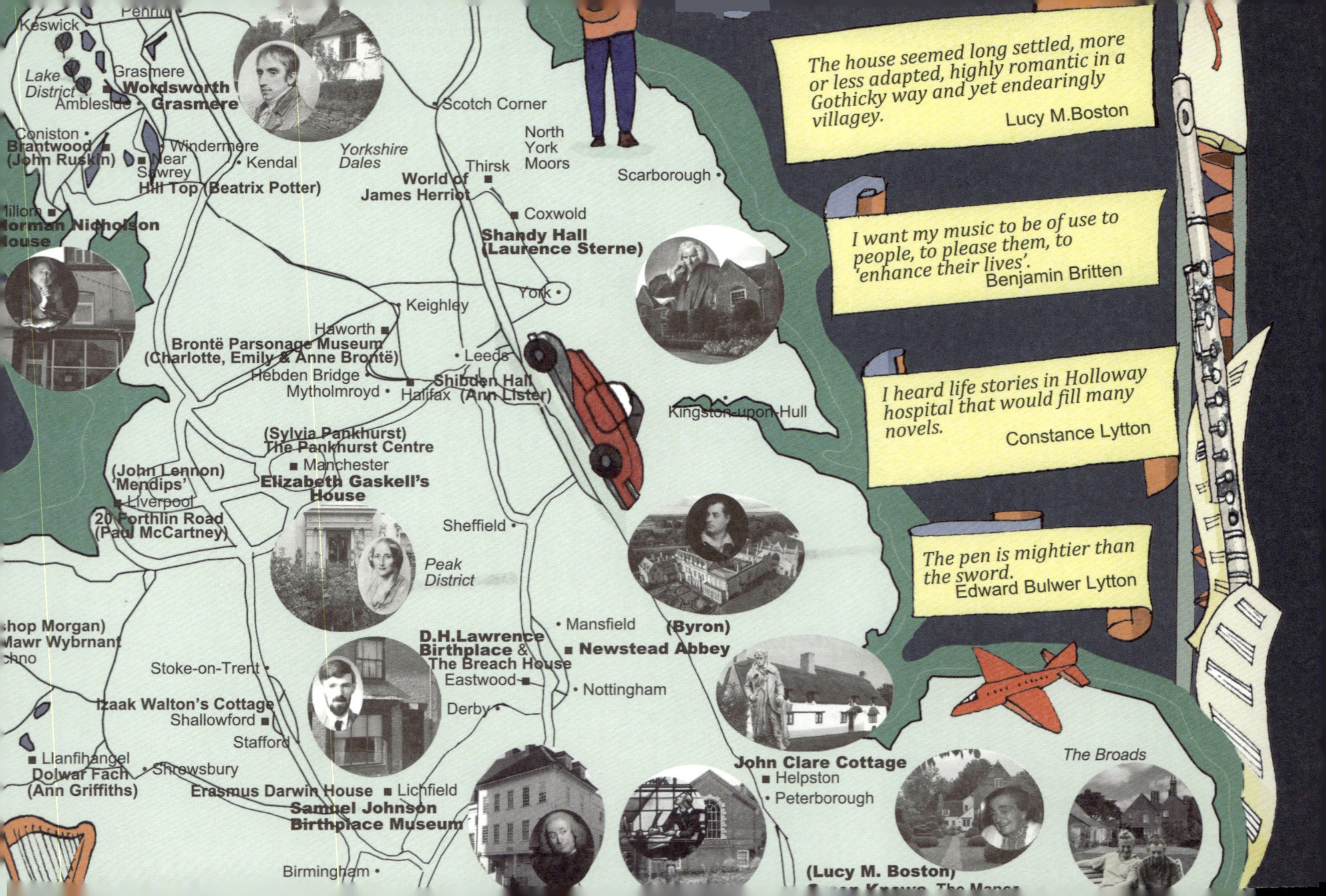

The house seemed long settled, more or less adapted, highly romantic in a Gothicky way and yet endearingly villagey.
Lucy M.Boston
I want my music to be of use to people, to please them, to 'enhance their lives'.
Benjamin Britten
I heard life stories in Holloway hospital that would fill many novels.
Constance Lytton
The pen is mightier than the sword.
Edward Bulwer Lytton
Keswick
Lake District
Grasmere
Wordsworth Grasmere
Ambleside
Coniston
Brantwood (John Ruskin)
Windermere
Near Sawrey
Hill Top (Beatrix Potter)
Kendal
Yorkshire Dales
Scotch Corner
North York Moors
Thirsk
World of James Herriot
Scarborough
Coxwold
Shandy Hall (Laurence Sterne)
York
Keighley
Haworth
Brontë Parsonage Museum (Charlotte, Emily & Anne Brontë)
Leeds
Hebden Bridge
Mytholmroyd
Halifax
Shibden Hall (Ann Lister)
Kingston-upon-Hull
(Sylvia Pankhurst) The Pankhurst Centre
Manchester
(John Lennon) 'Mendips'
Liverpool
20 Forthlin Road (Paul McCartney)
Elizabeth Gaskell's House
Sheffield
Peak District
Mansfield
(Byron)
Newstead Abbey
D.H.Lawrence Birthplace & The Breach House
Eastwood
Nottingham
Stoke-on-Trent
Izaak Walton's Cottage
Shallowford
Stafford
Derby
Llanfihangel
Dolwar Fach (Ann Griffiths)
Shrewsbury
Erasmus Darwin House
Lichfield
Samuel Johnson Birthplace Museum
Birmingham
John Clare Cottage
Helpston
Peterborough
(Lucy M. Boston)
The Broads

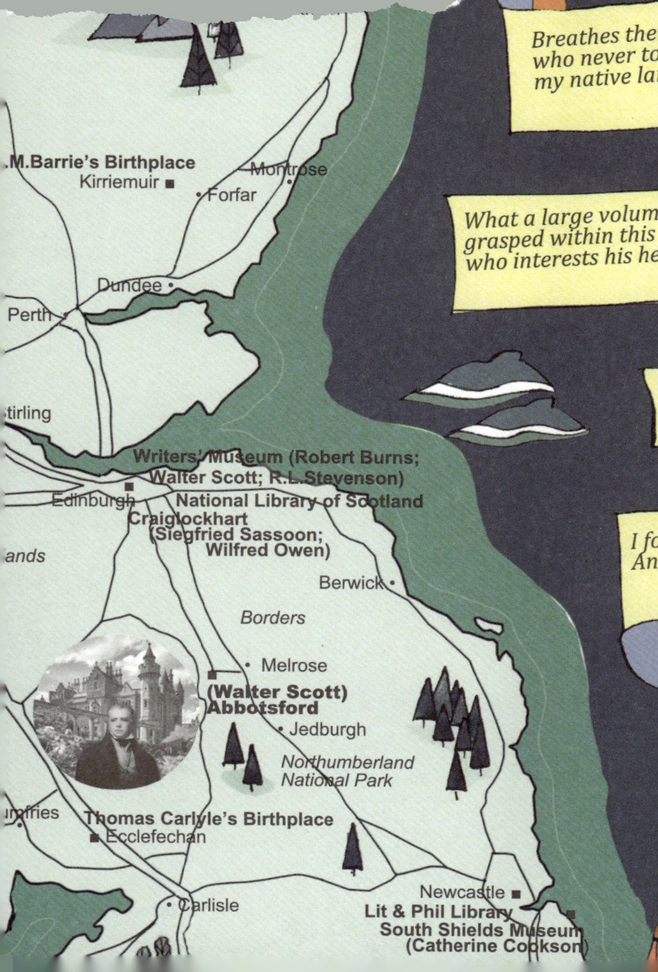

Breathes there the man with soul so dead,
who never to himself hath said, this is my own,
my native land!
Walter Scott

What a large volume of adventures may be grasped within this little span of life by him who interests his heart in every thing.
Laurence Sterne

Always laugh when you can, it is cheap medicine.
Byron

I found the poems in the fields,
And only wrote them down.
John Clare

What is written without effort is in general read without pleasure
Samuel Johnson

I would let moss grow on my eyes in prison rather than disobey God.
John Bunyan

the Madding Crowd 01305 262538 *NT*
Wilton House Wilton, Salisbury SP2 0BJ ***Mary Herbert*** (1561-1621), Countess of Pembroke, sister of poet Sir Philip Sidney, was a poet, literary patron and playwright. wiltonhouse.co.uk 01722 746714

WALES [W]

5 Cwmdonkin Drive Uplands, Swansea SA2 0RA Sensitively restored family home of ***Dylan Thomas*** (1914-53), Anglo-Welsh 'man of words'. Works include *Under Milk Wood* and 'Do not go gentle into that goodnight'. dylanthomasbirthplace.com 01792 472555
Cae'R Gors Caernarfon LL54 7EY Quarry-man's cottage, childhood home of Welsh language short story writer and novelist ***Kate Roberts*** (1891-1985) celebrated as Brenhines ein Llên 'Queen of our Literature'. 0300 025 6000 *CADW*
Dolwar Fach/Fechan Llanfihangel-Yng-Ngwynfa Home of Welsh poet and hymnologist ***Ann Griffiths*** (1776-1805) and place of pilgrimage for Welsh Methodists.
Dylan Thomas Boathouse Dylan's Walk, Laugharne, Carmarthen SA33 4SD ***Thomas***' inspirational final home with his writing hut. dylanthomasboathouse.com 01994427420
Lasynys Fawr Snowdonia National Park, Harlech LL46 2UT Birthplace of clergyman ***Ellis Wynne*** (1671-1734) author of the influential Welsh language masterpiece 'Visions of the Sleeping Bard'. 01766 780846
National Library of Wales Penglais Rd, Aberystwyth SY23 3BU Access to Welsh history and culture; exhibitions. library.wales
Strata Florida Abbey Ystrad Fflur, Ystrad Meurig SY25 6ES Ruins of the great Medieval abbey, burial place of Welsh princes and reputedly the poet ***Dafydd ap Gwilym*** under a yew in the churchyard. *CADW*
Tŷ Mawr Wybrnant Penmachno, Betws-y-Coed, Conwy, LL25 0HJ Farmhouse birthplace of Bishop ***William Morgan*** (1545-1604) the first translator of the Bible into Welsh, credited with saving the Welsh language. 01690760213 *NT*
Y Gangell Blaenycoed, Cynwyl Elfed, Carms, SA33 6EX Cottage birthplace of hymn writer and poet Rev.***Howell Elvet Lewis*** (1860-1953) aka ***Elfed***. (Not easily accessible)
Yr Ysgwrn Trawsfynydd, Blaenau Ffestiniog LL41 4UW Farmhouse home of soldier poet ***Ellis Evans***, Bard ***Hedd Wyn*** (1887-1917), killed in the First World War. Poet of the Black Chair. yrysgwrn.com 01766 772508

LONDON [L]

Benjamin Franklin House 36 Craven St, Charing Cross, London WC2N 5NF Only surviving house of ***Benjamin Franklin*** (1706-90), American Founding Father, writer, and polymath. benjaminfranklinhouse.org 0207 839 2006
British Library 96 Euston Rd, London NW1 2DB Holds numerous literary exhibitions bl.uk
Carlyle's House 24 Cheyne Row, Chelsea, London SW3 5HL Home of literary couple ***Thomas*** & ***Jane Carlyle***. Thomas wrote *The French Revolution*. 020 7352 7087 *NT*
Charles Dickens Museum 48-49 Doughty St, Holborn, London WC1N 2LX ***Charles Dickens*** (1812-70) became the most popular writer of the Victorian period and remains one of the most celebrated writers of all time. This Georgian terrace London house was his first family home, where he wrote *Oliver Twist*, *The Pickwick Papers* and *Nicholas Nickleby*. dickensmuseum.com 020 7405 2127
Dr Johnson's House 17 Gough Square, Holborn, London EC4A 3DE Atmospheric 17th century townhouse where wit, biographer, critic and essayist ***Samuel Johnson*** (1709-84) 'the second most-quoted Englishman' completed his Dictionary. A hidden gem off the Strand. 020 7353 3745 drjohnsonshouse.org
Freud Museum 20 Maresfield Gardens, London NW3 5SX The final home of ***Sigmund Freud*** (1856-1939) the founder of psychoanalysis, a theory of how the mind works and a method of helping people in mental distress. He wrote his last work, *An Outline of Psychoanalysis*, here, surrounded by his library and his beloved antiquities. freud.org.uk 020 7435 2002

Chartwell Mapleton Rd, Westerham TN16 1PS Home of ***Winston Churchill*** (1874-1965) wartime Prime Minister, soldier and writer. Works include *A History of the English-Speaking Peoples*. Nobel Laureate. 01732 868381 *NT*

Dickens House Museum 2 Victoria Parade, Broadstairs CT10 1QS Celebrates ***Dickens***' connection with Broadstairs. 01843 861232

Gad's Hill Place Higham, Rochester, Kent. ME3 7PA ***Charles Dickens***' final home, where he wrote his last 5 novels, including *Great Expectations*. gadshillplace.co.uk 01474 770034

Knole Sevenoaks TN15 0RP Jacobean mansion in the Sackville family since 1603 and inspiration for many writers including ***Vita*** and ***Eddy Sackville-West***, and ***Virginia Woolf***. 01732 462100 *NT*

Lamb House West St, Rye TN31 7ES 18th century house where ***Henry James*** (1843-1916) wrote some of his most acclaimed novels, including *Wings of the Dove*. Novelists ***E.F.Benson*** (1867-1940) and ***Rumer Godden*** (1907-98) also successively lived here. 01797 222909 *NT*

Leith Hill Place Dorking RH5 6LY Childhood home of English composer ***Ralph Vaughan Williams*** (1872-1958). Works include 'The Lark Ascending'. 01306 711685 *NT*

Monk's House Rodmell, Lewes BN7 3HF 16th century cottage home of Bloomsbury novelist, essayist and short story writer ***Virginia Woolf*** (1882-1941) and Leonard Woolf. Works include *Mrs Dalloway*, *To the Lighthouse*, *The Waves* 01273 474760 *NT*

Penshurst Place Tonbridge, Kent TN11 8DG 14th century manor house in the Sidney family since 1552, home of poet and soldier ***Sir Philip Sidney*** (1554-86) author of 'Arcadia'. penshurstplace.com 01892 870307

Smallhythe Place Smallhythe Rd, Tenterden TN30 7NG Suffragette playwright ***Christopher St John/Christabel Marshall*** (1871-1960) lived with theatre director Edith Craig and artist Tony Atwood in the grounds of actress Ellen Terry's 16th century cottage. 01580 762334 *NT*

SOUTH WEST ENGLAND [SW]

Beckford's Tower Lansdown Rd, Bath BA1 9BH The museum explores how the extreme wealth of eccentric novelist, art critic, architect ***William Beckford*** (1760-1844) was built on slavery. Beckford wrote the Gothic novel *Vathek*. beckfordstower.org.uk

Clouds Hill King George V Road, Bovington BH20 7NQ Final hideaway of archeologist, soldier and writer ***T.E.Lawrence*** (1888-1935), better known as Lawrence of Arabia. 01929 405616 *NT*

Coleridge Cottage 35 Lime St, Nether Stowey, Bridgwater TA5 1NQ 17th century cottage home of Romantic poet ***Samuel Taylor Coleridge*** (1772-1834) where he wrote 'Kubla Khan' and 'Rime of the Ancient Mariner' and, with Wordsworth, began the *Lyrical Ballads*. 01278 732662 *NT*

Greenway Galmpton, Devon TQ5 0ES Family holiday home of crime writer ***Agatha Christie*** (1890-1976) creator of Hercule Poirot and Miss Marple. 0344 249 1895 *NT*

Hardy's Cottage Higher Bockhampton, nr Dorchester, Dorset DT2 8QJ Birthplace of author ***Thomas Hardy*** 01305 262366 *NT*

Holst Birthplace Museum 4 Clarence Rd, Cheltenham GL52 2AY Time capsule of Regency and Victorian life with memorabilia of composer ***Gustav Holst*** (1874-1934) holstmuseum.org.uk 01242 524846

Jamaica Inn Bolventor, Launceston PL15 7TS Smugglers inn made famous by ***Daphne du Maurier***'s eponymous novel. Smugglers museum with room dedicated to du Maurier (1907-1989) jamaicainn.co.uk 01566 86250

Jane Austen Centre 40 Gay St, Bath BA1 2NT Small museum dedicated to the life and times of ***Jane Austen***. 01225 443000 janeausten.co.uk

Kelmscott Manor Lechlade GL7 3HJ 16th century manor house, inspirational retreat of writer, designer, socialist ***William Morris***. 01367 252486 sal.org.uk/kelmscott-manor

Max Gate Alington Ave, Dorchester DT1 2FN Self-designed house of celebrated novelist and poet ***Thomas Hardy*** (1840-1928) Works include *Tess of the d'Urbervilles* and *Far from*

A Map of LITERARY & MUSICAL HOUSES of Britain and Ireland
Orkney
George Mackay Brown)
John O'Groats
Wick

Lewis and Harris
Skye
The Highlands
Inverness

The Literary & Musical Houses

For current opening times, please consult houses' websites or phone in advance.

Please respect the privacy of privately owned properties and only visit during official opening times or by prior arrangement.

NT – National Trust property
NTS – National Trust for Scotland
EH – English Heritage property
CADW – Welsh Government historic environment service

With grateful thanks to **The LitHouses Group** www.lithouses.org for their support and the inspiration for this map.

SCOTLAND [Sc]

Abbotsford, Melrose TD6 9BQ ***Sir Walter Scott*** (1771-1832) novelist, historian and poet, transformed how the world saw Scotland and Scotland saw itself. Abbotsford is Scott's romance in lime and mortar, and a shrine to storytelling on the banks of the River Tweed. Works include 'The Lady of the Lake', *Ivanhoe*, *Waverley* and *Rob Roy*. 01896 752043 scottsabbotsford.com

Craiglockhart, Edinburgh Napier University, 219 Colinton Rd, EH14 1DJ Former war hospital where 1st World War poets ***Siegfried Sassoon*** (1886-1967) and ***Wilfred Owen*** (1893-1918) met. Houses The War Poets Collection. napier.ac.uk war poets collection

Gavin Maxwell Museum Author, naturalist, secret agent, ***Gavin Maxwell*** (1914-1969), wrote *Ring of Bright Water*, and lived on the tiny island of Eilean Bàn under the Skye Bridge. Book tours ahead. eileanban.org 01599 530040

J.M.Barrie's Birthplace 9 Brechin Rd, Kirriemuir DD8 4BX ***J.M.Barrie*** (1860-1937) world-famous playwright, novelist and creator of *Peter Pan*, was born in this traditional weaver's house. 01575 572646 *NTS*

Robert Burns Cottage & Museum 16 Alloway, Ayr KA7 4PY Poet ***Robert Burns*** (1759-96) Scotland's National Bard was born here. 01292 443700 *NTS*

Stromness Library 2-12 Victoria St, Stromness KW16 3AA houses a few personal belongings of Orkney poet ***George Mackay Brown*** (1921-1996) 01856 850907

The Boswell Quill & trail Old Churchyard, Auchinleck ***James Boswell*** (1740-95) traveller, diarist, advocate, Laird of Auchinleck, inventor of the modern biography, wrote *The Life of Samuel Johnson*. The bronze Boswell Quill stands outside the Boswell Mausoleum. Trails tell the story of the Boswells and Auchinleck. jamesboswell.scot

Thomas Carlyle's Birthplace Arched House, High St, Ecclefechan, Lockerbie DG11 3DG Childhood home of ***Thomas Carlyle*** (1795-1881), one of Britain's most influential thinkers, author of *On Heroes* and *The French Revolution* 01576 300666 *NTS*

Writers' Museum Lawnmarket, Lady Stair's Close, Edinburgh EH1 2PA celebrates and contains memorabilia of three giants of Scottish Literature: ***Robert Burns***; ***Walter Scott***; ***R.L.Stevenson*** 0131 529 4901 edinburghmuseums.org.uk

NORTHERN ENGLAND [N]

Brantwood East of Lake, Coniston LA21 8AD Museum, home of ***John Ruskin*** (1819-1900) influential art critic and radical social commentator, champion of Turner, the Pre-Raphaelites and Gothic architecture, wrote *The Stones of Venice* brantwood.org.uk 015394 41396

Brontë Parsonage Museum Church Street, Haworth, Keighley BD22 8DR Home of the novelists ***Charlotte*** (1816–1855), ***Emily*** (1818–1848) and ***Anne Brontë*** (1820–1849), authors of *Jane Eyre*, *Wuthering Heights* and *The Tenant of Wildfell Hall* respectively. 01535 642323 bronte.org.uk

Elizabeth Gaskell's House 84 Plymouth Grove, Manchester M13 9LW ***Elizabeth Gaskell*** (1810-65) wife, mother, writer, charity worker and traveller, most famous for writing *Cranford*, social realism novels, *Mary Barton*, *North and South* and the biography of her friend *Charlotte Brontë*. This beautifully restored middle-class Victorian home brings Manchester's literary heritage to life. 0161 273 2215 elizabethgaskellhouse.co.uk

Hill Top Near Sawrey, Ambleside LA22 0LF 17th century farmhouse home of beloved children's writer, artist and conservationist ***Beatrix Potter*** (1866-1943) author of *The Tale of Peter Rabbit*. 015394 36269 *NT*

Lit & Phil Library 23 Westgate Rd, Newcastle upon Tyne NE1 1SE Newcastle's exquisite secret free library, a hub of learning, enlightenment and the arts since 1825 litandphil.org.uk 0191 232 0192

'Mendips' 251 Menlove Ave, Liverpool L25 7SA Childhood home of Beatle, composer, peace activist and writer ***John Lennon*** (1948-80) 0344 800 1895 *NT*

Norman Nicholson House 14 St George's Terrace, Millom, Cumbria LA18 4DB ***Norman Nicholson*** (1914-87) poet and author whose work finds universal values in localness, community and the environment. This Victorian terraced house and former shop, his lifelong home, is now in private ownership but occasional Normal Nicholson Society events are held here. normannicholson.org

Pankhurst Centre 60–62 Nelson Street, Manchester Childhood home of ***Sylvia Pankhurst*** (1882-1960) suffragette, socialist, radical activist, artist and writer, daughter of Emmeline Pankhurst. Birthplace of the suffragette movement (WSPU). 0161 273 5673 pankhursttrust.org

Shandy Hall Coxwold, York YO61 4AD ***Laurence Sterne*** (1713-68) vicar in Coxwold, North Yorkshire, wrote *The Life and Opinions of Tristram Shandy, Gentleman* and *A Sentimental Journey*. Dating from 1430, the house where Sterne lived and wrote is a lived-in museum and contains a collection of Sterne-related books and works of art. laurencesternetrust.org.uk 01347 868465

Shibden Hall Lister's Rd, Halifax HX3 6XG Dating from 1420, home of the noted diarist ***Ann Lister*** (1791-1840) 01422 352246 museums.calderdale.gov.uk

South Shields Museum Ocean Rd, South Shields NE33 2JA Memorabilia from the life of ***Catherine Cookson*** (1906-98), one of the most widely read British authors, including her Emmy award for *The Black Velvet Gown*. 0191 211 5599 southshieldsmuseum.org.uk

20 Forthlin Road Liverpool L18 9TN Childhood home of Beatle and musician ***Paul McCartney*** (b.1942) 0844 800 4791 *NT*

World of James Herriot 23 Kirkgate, Thirsk YO7 1PL Home of the vet and author ***James Herriot*** (1916-95) worldofjamesherriot.com 01845 524234

Wordsworth Grasmere Dove Cottage, Grasmere, Cumbria LA22 9SH ***William Wordsworth*** (1770-1850) poet, celebrated our relationship with nature and the importance of taking time to appreciate the beauty of the natural world. His poetry is synonymous with the unique landscape of the English Lake District. 015394 35544 wordsworth.org.uk

MIDLANDS [M]

Breach House 28 Garden Rd, Eastwood, Nottingham NG16 3FW ***D.H.Lawrence***'s childhood home; a typical miner's cottage of the late 1800s. thebreachhouse.co.uk

D.H.Lawrence Birthplace 8a Victoria Street, Eastwood, Nottingham NG16 3AW Poet and novelist ***D.H.Lawrence*** (1885-1930), the son of a coalminer, developed into one of the most influential writers of the 20th century. Novels include *Sons & Lovers*, *Women in Love* and *Lady Chatterley's Lover*. The birthplace is an authentically restored miner's cottage. lleisure.co.uk/d-h-lawrence-birthplace-museum

Erasmus Darwin House Beacon St, Lichfield WS13 7AD Family home of ***Erasmus Darwin*** (1731-1802) doctor, inventor, poet and grandfather of evolutionary biologist Charles Darwin. 01543 306260 erasmusdarwin.org

Izaak Walton's Cottage Worston Lane, Shallowford, Stone ST15 0PA ***Izaak Walton*** (1593-1683) influential biographer, writer, angler and royalist, wrote *The Compleat Angler* and *Walton's Lives* 01785 760278

How easily & cleverly do I write just now! Words come skipping to me like lambs upon Moffat-hill.
James Boswell
Come forth into the light of things, Let Nature be your teacher.
William Wordsworth
I'm grateful for roots, I don't think I would have developed as a poet anywhere else.
Norman Nicholson
We've got a house...it certainly is a beauty... I must try and make the house give as much pleasure to others as I can.
Elizabeth Gaskell
We've got to live, no matter how many skies have fallen.
D.H.Lawrence
Loch Lomond and the Trossachs
Jura
Glasgow
Kilmarnock
Ayr
Alloway
Robert Burns Cottage & Museum
Glenveagh National Park
Londonderry/Derry

HomePlace
(Seamus Heaney)
Maghera
Bellaghy
Donegal
Omagh
Lough Neagh
Belfast
(Eva Gore-Booth, Constance Markievicz, Yeats)
Lissadell House
Enniskillen
Armagh
Morne Mountains
Isle of Man
Sligo
Yeats Building
Newry
Ballycroy National Park
Cavan
Inniskeen
Patrick Kavanagh Centre
Dundalk
Knock
Longford
Edgeworthstown
Maria Edgeworth Centre
Kells
Slane
Drogheda
Francis Ledwidge Cottage Museum
Mullingar
Athlone
Galway
Dublin
Holyhead
Anglesey
Dun Laoghaire
James Joyce Tower and Museum
Caernarf
Cae'R Gors
(Kate Roberts)
Thoor Ballylee (W.B.Yeats)
Kiltartan Gregory Museum
Wicklow Mountains
Coole Park
Gort
Portlaoise
Wicklow
Lasynys Fawr
(Ellis Wynne)
Harlech
Limerick
Cashel
Cardigan Bay

Newstead Abbey Newstead Village NG15 8GE Former 12th century monastic abbey, ancestral home of celebrated/notorious Romantic poet ***Lord George Gordon Byron*** (1788-1824), author of *Don Juan* and *Childe Harold's Pilgrimage*. 0115 8763100 newsteadabbey.org.uk

Samuel Johnson Birthplace Museum Breadmarket St, Lichfield WS13 6LG ***Samuel Johnson*** (1709-84) the 18th century's most famous man of letters, known for his *Dictionary of the English Language*. This Grade 1 listed townhouse is a museum dedicated to his life and times. 01543 264972 samueljohnsonbirthplace.org.uk

Shakespeare's Birthplace Henley St, Stratford-upon-Avon CV37 6QW ***William Shakespeare*** (1564-1616) Britain's most renowned playwright/poet whose legacy to literature and understanding of human nature in plays such as *Hamlet*, *King Lear* and *Romeo and Juliet* is unsurpassed. The Birthplace is one of several houses cared for by the Shakespeare Birthplace Trust around Stratford. 01789 204016 shakespeare.org.uk

The Firs: Elgar's Birthplace Crown East Lane, Lower Broadheath, Worcester WR2 6RH ***Edward Elgar*** (1857-1934) from humble origins became one of Britain's foremost composers. Works include his cello concerto and *Enigma Variations*. 01905 333330 *NT*

EASTERN ENGLAND [E]

Cowper & Newton Museum Olney MK46 4AJ Home of poet (quoted by Martin Luther-King Jnr), hymn writer and translator of Homer, ***William Cowper*** (1731-1800), friend of preacher and slave trader turned abolitionist ***John Newton*** (1725-1807) author of the hymn 'Amazing Grace'. 01234 711516 cowperandnewtonmuseum.org.uk

Green Knowe, The Manor, High Street, Hemingford Grey, Huntingdon PE28 9BN ***Lucy M. Boston*** (1892-1990) novelist, gardener, artist, published entirely after she turned 60. She based her popular children's light fantasies on this stone moated manor house, built around 1130, beside the River Great Ouse. greenknowe.co.uk 01480 463134

John Bunyan Museum & Library, Bunyan Meeting, Mill Street, Bedford, MK40 3EU ***John Bunyan*** (1628-88) renowned dissenting preacher, prisoner of conscience, and author of over 60 books, including *The Pilgrim's Progress*, lived through the intense social, civil, and religious upheaval of the English Civil War, Commonwealth and Restoration. He was imprisoned for 12 years. Bunyanmeeting.co.uk

John Clare Cottage Woodgate, Helpston, Peterborough PE6 7ED Birthplace of poet ***John Clare*** (1793-1864). Son of a farm labourer, he celebrated the English countryside in works such as *The Shepherd's Calendar,* and is now seen as a major 19th century Romantic poet. clarecottage.org 01733 253330

Knebworth House Stevenage SG1 2AX ***Edward Bulwer Lytton*** (1803-73) writer and politician, best-selling author of the 1830s, wrote novels, poetry and plays in genres from historical to the occult. His wife (separated) ***Rosina Doyle Wheeler*** (1802-1882) was a novelist and essayist; their granddaughter, the suffragette ***Constance Lytton*** wrote the influential *Prisons and Prisoners* about her experiences. In the Lytton family since 1490, this crenelated high gothic Victorian fantasy conceals a Tudor brick manor. knebworthhouse.com 01438 812661

Shaw's Corner Bibbs Hall Lane, Welwyn AL6 9BX Country home of outspoken, internationally famous playwright, socialist, feminist, pacifist ***George Bernard Shaw*** (1856-1950) 01438 820307 *NT*

The Red House, Aldeburgh, IP15 5PZ ***Benjamin Britten*** (1913-76), one of the greatest composers of the 20th century, works including *Peter Grimes* and *War Requiem*, lived here with life partner and inspiration, the singer ***Peter Pears*** (1910-86). A large 17th century farmhouse in the grounds of a five-acre garden, furnished with an eclectic mix of traditional and contemporary styles and an outstanding art collection. brittenpearsarts.org 01728 451700

SOUTH OF ENGLAND [S]

Blake's Cottage Blakes Rd, Bognor Regis PO22 7EE House of visionary poet, artist and radical ***William Blake*** (1757-1827) author of 'Jerusalem'. Renovation plan. blakecottage.org

Charles Dickens' Birthplace Museum 393 Old Commercial Rd, Portsmouth PO1 4QL charlesdickensbirthplace.co.uk 023 9282 7261

Chawton House, Chawton, Alton GU34 1SJ ***Jane Austen*** knew this Elizabethan manor owned by her brother Edward as 'the Great House'. Filled with Austen and Knight family heirlooms, it houses a major collection of early women's writing, including works by actress ***Mary Robinson*** (1758-1800) and botanist ***Elizabeth Blackwell*** (1700-1758). chawtonhouse.org 01420 541010

Farringford Freshwater, Isle of Wight Home of the Poet Laureate, ***Alfred Tennyson*** (1809-92) author of *Idylls of the King* and 'In Memoriam' 01983 752500 farringford.co.uk

Gilbert White's House The Wakes, High Street, Selborne GU34 3JH In this 17th century house and garden ***Gilbert White*** (1720-93) father of ecology, pioneering naturalist and nature writer, revolutionised the way we look at the natural world, paving the way for future scientists. 01420 511 275 gilbertwhiteshouse.org.uk

Guildford Museum Quarry St, Guildford GU1 3SX Memorabilia of mathematician, classicist and author of the *Alice in Wonderland* books. ***Lewis Carroll/ Charles Dodgson*** (1832-98), who often stayed with his sisters in Guildford. 01483 444751

Hughenden Manor Hughenden HP14 4LA Impressive country home of ***Benjamin Disraeli*** (1804-81) Conservative prime minister and novelist. 01494 755573 *NT*

Jane Austen's House Winchester Road, Chawton, Hampshire GU34 1SD where ***Jane Austen*** (1775-1817) lived her last eight years writing, revising and publishing her world-famous, much loved novels, including *Pride and Prejudice*, *Emma*, *Sense and Sensibility*. 01420 83262 janeaustens.house

Milton's Cottage 21 Deanway, Chalfont Saint Giles HP8 4JH Discover the extraordinary story of ***John Milton*** (1608-74) – the visionary poet, parliamentarian and pamphleteer – at his only surviving residence. Explore the rural cottage and garden where he came to escape the Great Plague of 1665 and complete his epic masterpiece, *Paradise Lost*. 01494 872313 miltonscottage.org

Roald Dahl Museum & Story Centre, 81-83 High St, Great Missenden HP16 0AL Children's and short story writer, spy and inventor, ***Roald Dahl*** (1916-90) lived here for 36 years. Works include *The Witches*, *Charlie and the Chocolate Factory*, and *The BFG*. roalddahl.com 01494 892192

The Kilns Headington, Oxford OX3 8JD Home of English Literature academic and lay theologian ***Clive Staples Lewis*** (1898-1963) author of the world-famous Narnia books. Tours by appointment only. cslewis.org

GREATER LONDON [GL]

Down House Luxted Rd, Downe, Orpington BR6 7JT Home of scientist and naturalist ***Charles Darwin*** (1809-82), internationally renowned for his contributions to the theory of evolution in *On the Origin of Species*. 0370 333 1181 *EH*

The Red House Bexleyheath DA6 8JF Arts and Crafts home of designer and socialist author ***William Morris*** (1834-96). 0208 303 6359 *NT*

Strawberry Hill House Twickenham TW1 4ST Gothic-style home of **Horace Walpole** (1717-97) who invented the gothic novel with *The Castle of Otranto*. 0208 744 1241 strawberryhillhouse.org.uk

William Morris Gallery Lloyd Park, Forest Rd, Walthamstow, London E17 4PP Childhood home and museum of the life and work of inspiring socialist designer ***William Morris***. wmgallery.org.uk 020 84964390

SOUTH EAST ENGLAND [SE]

Bateman's, Burwash TN19 7DS 17th century house, home of poet, journalist, novelist, short story writer and Nobel Laureate ***Rudyard Kipling*** (1865-1936). Works include *The Jungle Book* and *Kim*. 01435 882302 *NT*

Mallow
Waterford
Rosslaire Harbour
Killarney National Park
Cork
Frank O'Connor House
(Edith Somerville) Drishane House
Castletownshend
Aberystwyth
National Library of Wales
Cardigan
Fishguard
Y Gangell
Blænycoed
St Clears
Carmarthen
Dylan Thomas Boathouse
Lougharne
(Dylan Thomas)
5 Cwmdonkin Drive
...love all, trust a few,
Do wrong to none...
Shakespeare
There is in the short story ... something we do not often find in the novel – an intense awareness of human loneliness.
Frank O'Connor
My first November at Shreelane was composed of weather of which my friend Flurry Knox remarked that you wouldn't meet a Christian out of doors, unless it was a snipe or a dispensary doctor.
Edith Somerville
The parish I live in is a very abrupt, uneven country, full of hills and woods, and therefore full of birds.
Gilbert White
Bodmin Moor
Bolventor
Jamaica Inn
(Daphne du Maurier)
Bodmin
Newquay
Plymouth
Penzance

Dublin
Dublin Writers Museum
James Joyce Centre
Street
Gardiner
Connolly Station
Parnell
O'Connell Street
Talbot St
St
Henry Street
GPO
Street
Abbey
River Liffey
Temple Bar
Pearse Street
Trinity College, Book of Kells & Old Library
Grafton St
Kildare St
National Library
Merrion Square
St Patrick's Cathedral
St Stephen's Green
South
Museum of Literature Ireland
A good book is the precious life-blood of a master spirit, embalmed and treasured up on purpose to a life beyond life.
John Milton
I went up to the Great House between 3&4, & dawdled away an hour very comfortably.
Jane Austen
Why are the graces of feminine beauty to be constituted emblems of a debilitated mind?
Mary Robinson

Discover the homes of British and Irish writers and musicians

Explore the houses open to the public that were once home to some of the finest writers and musicians in Britain and Ireland. This illustrated map shows where to find the birthplaces, lifelong residences or retreats where artists found peace and inspiration.

- Hand-drawn map featuring illustrations of authors and their houses
- Short commentaries on the authors with location and contact details of the houses
- Museums and libraries with permanent exhibitions on local authors, or major exhibitions on literary subjects
- Excerpts from the works of writers from the 16th to the 20th century

This map is also available as a limited edition giclée print. For information on prints and on other maps in this series, visit www.literarymapcompany.com

First published in the United Kingdom in 2021 by The Literary Map Company.

Printed in Histon, Cambridge by Print Out

ISBN 978-1-9160743-9-2

9 781916 074392 >

THE LITERARY MAP COMPANY